MACK MODEL L FIRE TRUCKS 1940 THROUGH 1954

PHOTO ARCHIVE

Iconografix Inc. exists to preserve history through the publication of notable photographic archives and the list of titles under the Iconografix imprint is constantly growing. Transportation enthusiasts should be on the Iconografix mailing list and are invited to write and ask for a catalog, free of charge.

Authors and editors in the field of transportation history are invited to contact the Editorial Department at Iconografix, Inc., PO Box 446, Hudson WI 54016. We require a minimum of 120 photographs per subject. We prefer subjects narrow in focus, e.g., a specific model, railroad, or racing venue. Photographs must be of high-quality, suited to large format reproduction.

MACK MODEL L FIRE TRUCKS 1940 THROUGH 1954

PHOTO ARCHIVE

Photographs from the
Mack Trucks Historical Museum Archives

Edited with introduction by
Harvey Eckart

Iconografix
Photo Archive Series

Iconografix
PO Box 446
Hudson, Wisconsin 54016 USA

Library of Congress Card Number 98-71286

ISBN 1-882256-86-7

98 99 00 01 02 03 04 5 4 3 2 1

Printed in the United States of America
Cover and book design by Shawn Glidden

Cover Photo: FDNY rescue truck.

PREFACE

The histories of machines and mechanical gadgets are contained in the books, journals, correspondence, and personal papers stored in libraries and archives throughout the world. Written in tens of languages, covering thousands of subjects, the stories are recorded in millions of words.

Words are powerful. Yet, the impact of a single image, a photograph or an illustration, often relates more than dozens of pages of text. Fortunately, many of the libraries and archives that house the words also preserve the images.

In the *Photo Archive Series*, Iconografix reproduces photographs and illustrations selected from public and private collections. The images are chosen to tell a story—to capture the character of their subject. Reproduced as found, they are accompanied by the captions made available by the archive.

The Iconografix *Photo Archive Series* is dedicated to young and old alike, the enthusiast, the collector and anyone who, like us, is fascinated by "things" mechanical.

Bibliography

Montville, John B. MACK, Aztec Corporation, Tucson, Arizona 1979

Eckart, Harvey MACK FIRE APPARATUS, A PICTORIAL HISTORY. The Engine House, Middletown, New York, 1990

Calderone, John FDNY MACK L MODEL APPARATUS, Fire Apparatus Journal Publications, Inc., Staten Island, New York, 1992

Acknowledgments

The photographs appearing in this book were made available by the Mack Trucks Historical Museum, the private collections of Old Dominion Historical Fire Society, and the editor.

75S2315 1948 World famous FDNY Rescue Co. No I took delivery of this very impressive rescue rig in 1948. The chassis is an LF model with a 510 cubic inch thermodyne motor. The body and compartments were designed by FDNY members, and built by Approved. The cost was $26,600. External features shown are dual mirrors, fender mounted turn signals, a grille guard, a Mars light, and a Sireno siren.

INTRODUCTION

In 1940 Mack introduced its L series heavy duty truck line. The fire truck version, with its extremely attractive styling and superior road and pumping performance, was destined to become the heavy duty "flagship" in fire departments across America, from New York, Boston, and Baltimore, to Los Angeles, and Chicago.

Civilian production of both trucks and fire apparatus was curtailed during the World War II years (from 1942 until late 1944) except for essential users.

During the early part of the war, Mack fire apparatus production was moved from Allentown, Pennsylvania to Long Island City, New York. The chassis continued to be built in Allentown. After construction, they were driven to Long Island City, where the bodies were built and final assembly was completed.

By 1948, the production lines of all manufacturers including Mack were turning out record numbers of apparatus to fulfill the urgent requirements of the fire service due to limited production during the war years. By 1948, these vehicles were once again works of beauty, with the return of chrome and gold leaf in generous amounts. The gleaming chrome radiator shell of the L model Mack, along with the distinctive motor sound, made it instantly recognizable whether sitting in the station or working on the fire ground.

During the reign of the L model, Mack also produced substantial numbers of light and medium duty apparatus based on the E series chassis. In the early 1950s, the A series replaced the E series and stayed in production until 1954. In the fall of 1954 the B model fire apparatus replaced both the L and A series apparatus.

The L model apparatus production consisted mainly of pumpers and aerial units, although a complete line of city service ladder trucks, quads, quints, hose wagons, and specialized units were also available. Mack advertising proudly touted the fact that Mack built such vital parts as motors, clutches, transmissions, rear axles, brakes, steering gears, frames, sheet metal parts, and bodies in its own plants. Hale pumps were used primarily with Mack-designed and built pressure governors.

Vendor motors made by Hercules and Hall Scott were installed in a small number of higher capacity units, but the vast majority were powered by Mack thermodyne motors of 611 or 707 cubic inch displacement. The 707A and B dual ignition gasoline motor earned a solid reputation for high performance and dependability.

A few rotary gear pumps were installed, but Hale centrifugal pumps of two or four stages dominated. Sizes ranged from 750 to 2,000GPM, with 750 being the most popular. Five-speed transmissions were used initially, but most later units used a standard four-speed direct unit. Hydraulic brakes were standard, but unlike most competitors of the time, Mack offered air brakes which were specified in considerable numbers. Cab choices included open, semi open, closed, streamlined, and sedan cabs.

Mack L series apparatus were built to conform with the National Fire Protection Association (NFPA) standards. In addition to the minimum requirements, a wide range of options were available. The photos to follow are arranged by serial number in chronological order (for the most part) by types, and illustrate the wide variety and diversity of L model units.

In 1951, fire apparatus production moved back to Allentown, Pennsylvania. The last L model commercial trucks rolled off the line in 1962 after over 30,000 had been built. The L model fire apparatus production ended in 1954 with 1,453 units built.

Both the L model commercial and fire apparatus series are revered by many as being the best "top of the line" series ever built by Mack. They truly were classic vehicles in a classic era. Fortunately, many examples have been restored and preserved for future generations to enjoy. In a few cases, some are still answering fire calls today, the ultimate tribute to the L models' reputation for dependability.

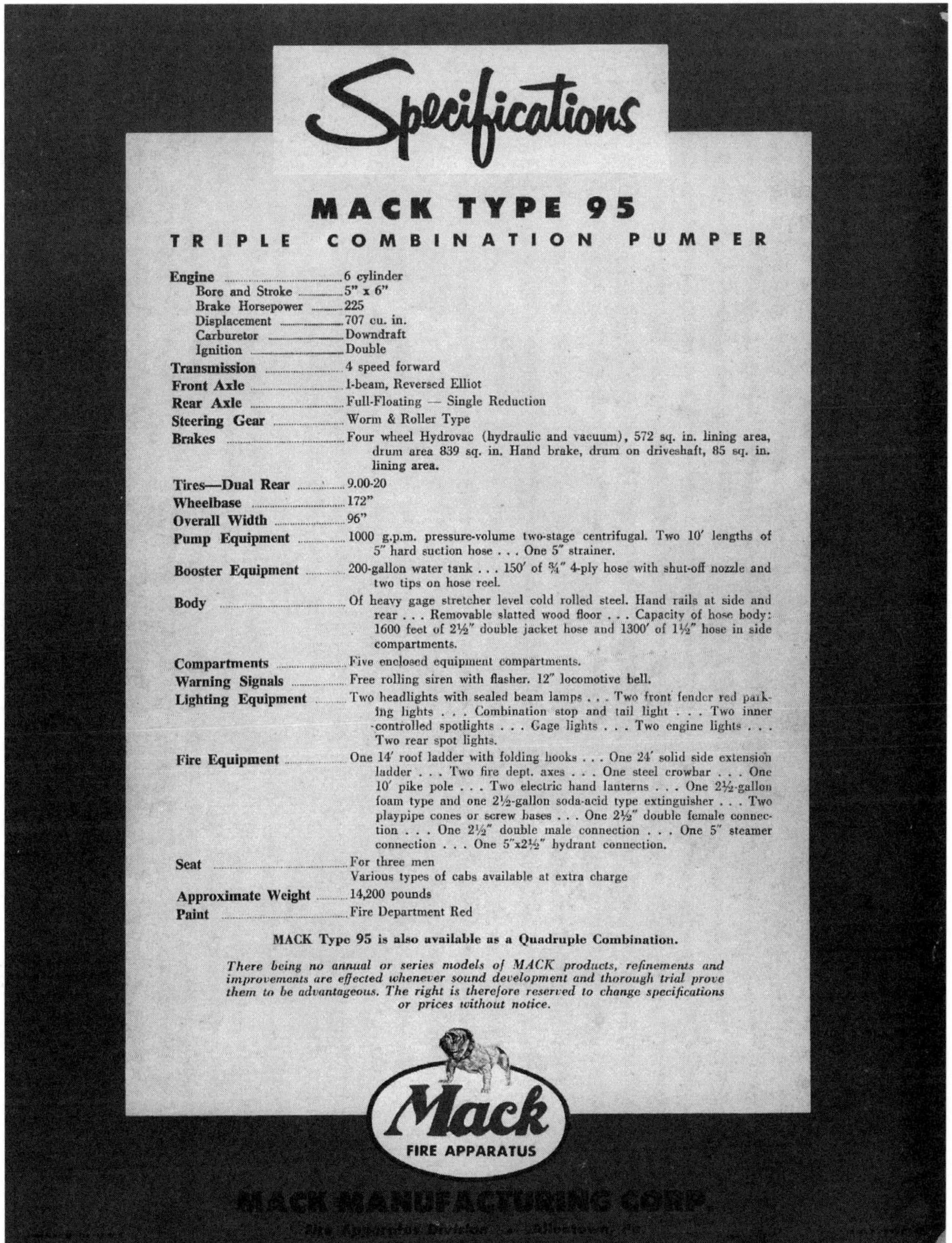

Specifications

MACK TYPE 95

TRIPLE COMBINATION PUMPER

Engine	6 cylinder
Bore and Stroke	5" x 6"
Brake Horsepower	225
Displacement	707 cu. in.
Carburetor	Downdraft
Ignition	Double
Transmission	4 speed forward
Front Axle	I-beam, Reversed Elliot
Rear Axle	Full-Floating — Single Reduction
Steering Gear	Worm & Roller Type
Brakes	Four wheel Hydrovac (hydraulic and vacuum), 572 sq. in. lining area, drum area 839 sq. in. Hand brake, drum on driveshaft, 85 sq. in. lining area.
Tires—Dual Rear	9.00-20
Wheelbase	172"
Overall Width	96"
Pump Equipment	1000 g.p.m. pressure-volume two-stage centrifugal. Two 10′ lengths of 5″ hard suction hose . . . One 5″ strainer.
Booster Equipment	200-gallon water tank . . . 150′ of ¾″ 4-ply hose with shut-off nozzle and two tips on hose reel.
Body	Of heavy gage stretcher level cold rolled steel. Hand rails at side and rear . . . Removable slatted wood floor . . . Capacity of hose body: 1600 feet of 2½″ double jacket hose and 1300′ of 1½″ hose in side compartments.
Compartments	Five enclosed equipment compartments.
Warning Signals	Free rolling siren with flasher. 12″ locomotive bell.
Lighting Equipment	Two headlights with sealed beam lamps . . . Two front fender red parking lights . . . Combination stop and tail light . . . Two inner-controlled spotlights . . . Gage lights . . . Two engine lights . . . Two rear spot lights.
Fire Equipment	One 14′ roof ladder with folding hooks . . . One 24′ solid side extension ladder . . . Two fire dept. axes . . . One steel crowbar . . . One 10′ pike pole . . . Two electric hand lanterns . . . One 2½-gallon foam type and one 2½-gallon soda-acid type extinguisher . . . Two playpipe cones or screw bases . . . One 2½″ double female connection . . . One 2½″ double male connection . . . One 5″ steamer connection . . . One 5″x2½″ hydrant connection.
Seat	For three men Various types of cabs available at extra charge
Approximate Weight	14,200 pounds
Paint	Fire Department Red

MACK Type 95 is also available as a Quadruple Combination.

There being no annual or series models of MACK products, refinements and improvements are effected whenever sound development and thorough trial prove them to be advantageous. The right is therefore reserved to change specifications or prices without notice.

Mack
FIRE APPARATUS

MACK MANUFACTURING CORP.
Fire Apparatus Division . . . Allentown, Pa.

Type 95 chassis details and standard fire equipment supplied, from a 1953 sales brochure.

The vast majority of L model Mack fire apparatus was completed at the Mack Long Island City, New York plant. The plant was built in 1925. Both the plant and parking lot took up an entire city block. The plant was made of reinforced concrete construction, having a floor area of 277,000 square feet. At one time or another, the plant was home to a rebuilding facility, body shop, service department, engineering department, and New York home office.

Fire apparatus manufacturing was done at Long Island City from the early part of World War II until 1951, when production was transferred back to Allentown. The plant was sold to CBS Columbia in 1952. Fire apparatus chassis were built in Allentown and driven to Long Island City. The extensive machine and body shop facilities at Long Island City were ideally suited to fire apparatus production. In addition, very skilled craftsman worked at the plant. Even a cursory look at an L model sedan cab unit will reveal the intricate metal work and craftsmanship of these units. The August 1949 photo on page 12 shows an open cab L model pumper exiting the parking lot with the four story plant in the background. The photo on page 13 shows both L and E model apparatus being assembled on the shop floor. Not much chrome is evident as the photo was taken in April, 1946.

TYPE 80

The modern era type 80 was introduced in 1938 in the E series. The L series type 80 was produced from 1940 through 1944 with 141 units built. The first two 80LSs went to Fort Worth, Texas. The final 80LS, serial number 80LS1202 was delivered on December 29, 1944 to Lockheed Aircraft Corp. in Burbank, California.

The type 80 used a 611 cubic inch overhead valve thermodyne motor with dual ignition. The standard pump was a Hale, two stage centrifugal of 750GPM capacity.

8OLSl053 1940 An early open cab pumper for Prospect Park, New Jersey. Non-sealed beam headlights, a full windshield header, and rubber covered running boards were characteristics of the very early units.

80LS1053 A searchlight and bell were standard. The CO2 extinguisher, double coat and boot rails, and soft suction hose holder were options.

80LS1067 1941 A bare chassis with pump and gauges mounted. This unit was delivered to the Penna Railroad in Altoona, Pennsylvania, after completion.

80LS1057 1941 L model sedan cabs were popular from the very beginning, as pictured on this early delivery to Norfolk, Virginia.

80LS1057 Several sedan cab configurations were used. The 3-door version (as shown), was the most popular. Later units had a single rear window rather than the dual windows shown on this pumper.

80LS1058 1941 This fender skirted, two-tone green pumper was delivered to the Los Angeles County Forestry and Fish and Game Department which was responsible for fire protection in rural Los Angeles County. Macks were the preferred builder prior to Crown's debut in the early 1950s.

80LS1069 1941 Sealed beam headlights are shown on this white pumper delivered to Ogden, Utah. The amount and placement of ladders deviated from standard.

80LS1078 1941 This sedan cab for North Adams, Massachusetts, is distinguished by its sharp, two-tone paint job.

80LS1110 1942 This blue and cream pumper went to Lewistown, Pennsylvania. Both this pumper and a 1938 Ahrens Fox BT stablemate served until 1974.

80LS1115 1943 This wartime delivery to Dubois, Pennsylvania, was completely devoid of chrome. This was designated as a streamlined cab in Mack sales literature.

80LS1115 There is no denying this cab style was attractive. In most cases, the area behind the seat was used for storage.

80LS1154 1943 This semi cab Quad for Canton, Ohio, was long and sleek. Popular at the time, the quadruple combination provided a pump, water tank, hose, and long ladders.

80LS1174 1943 Mack records do not list any reference to the trailer part of this tillered aerial delivered to Santa Barbara, California. It was not uncommon for the Los Angeles Branch or a West Coast body builder to complete final assembly of a unit. This photo is believed to have been taken in California.

80LS1171 1943 This wartime delivery to Kingston, Pennsylvania, is unique. It was equipped with a Hale CBMX 750 GPM rotary pump. Rotary pumps were extremely rare on L model apparatus.

80LS1171 The shift plate shown is a tip-off that a rotary pump was installed, since there is no "prime" position indicated, only "road" and "pump."

80LS1183 1944 One of the last type 80s delivered was this semi cab for Milton, Pennsylvania. Few options were installed except for the extra hard suction hose tray.

This cutaway pump display stand shows the standard Hale two-stage centrifugal pump, as used on type 80 units.

TYPE 85

The type 85LS was produced from 1940 through 1954 and was the best selling L type with 726 produced. The first two were delivered to San Antonio, Texas in 1940. The last 85LS, serial number 85LS1732 was delivered to Scotia, New York, on December 31, 1954.

The type 85 used a 707 cubic inch overhead valve thermodyne motor with dual ignition. The early version used an updraft carburetor, with the later 707A and B versions using a downdraft unit. This motor had a distinctive sound and became highly regarded for its performance and reliability.

A 750GPM Hale, two stage centrifugal pump was standard on the 85LS.

85LS1002 1940 San Antonio, Texas, received the first LS85s off the line. The decorative scrolls on the fenders were an option.

85LS1003 1941 The third LS85 off the line was this massive pumper for Boardman Township in Youngstown, Ohio. It had both a searchlight and spotlights, plus an optional warning light arrangement above the siren.

85LS1003 This pumper is considerably longer and larger than standard, with a cavernous hose bed.

85LS1071 1946 This sedan cab for Manchester, New Hampshire, lacks a center side window. It is equipped with an extra length of hard suction hose.

85LS1071 On this style cab, the rear seat faces rearward, and is entered via a walkway between the hose beds. Only a single tail and stop light was standard at the time.

85LS1096 1946 When a four-door sedan cab was specified, the ladders were normally relocated to a chute under the hose bed, or in an overhead rack (as shown). This unit was delivered to Melvindale, Michigan.

85LS1096 The rear seat area was quite spacious. The open hinged door provided access to the primer oil tank.

85LS1115 1948 This black and white pumper was delivered to Riverdale, Maryland. It was used as the model for a modern fire engine on a stamp issued in 1948, honoring 300 years of volunteer fire fighting in America.

85LS1115 The metal enclosure around the pipe to the high capacity deck gun shows the attention to small details in the construction of bodies.

85LS1140 1947 Always a loyal, high volume customer, New York City used types 19, 85, and 21 L models. This unit is one of 20 LS85 pumpers delivered in 1947 from two separate orders of ten each. Each pumper cost $15,113 and served for over 20 years. The Macks were joined by 20 pumpers from Ward LaFrance in 1946, and 20 from American LaFrance in 1947. What a two-year period that was!

85LS1140 New York trademarks were the rear "subway straps," a pompier ladder, a giant deck gun, and prominent nozzle tips.

85LS1140 Shown is a large life net compartment, the hard suction hose, and pompier ladders carried on both sides of the rig.

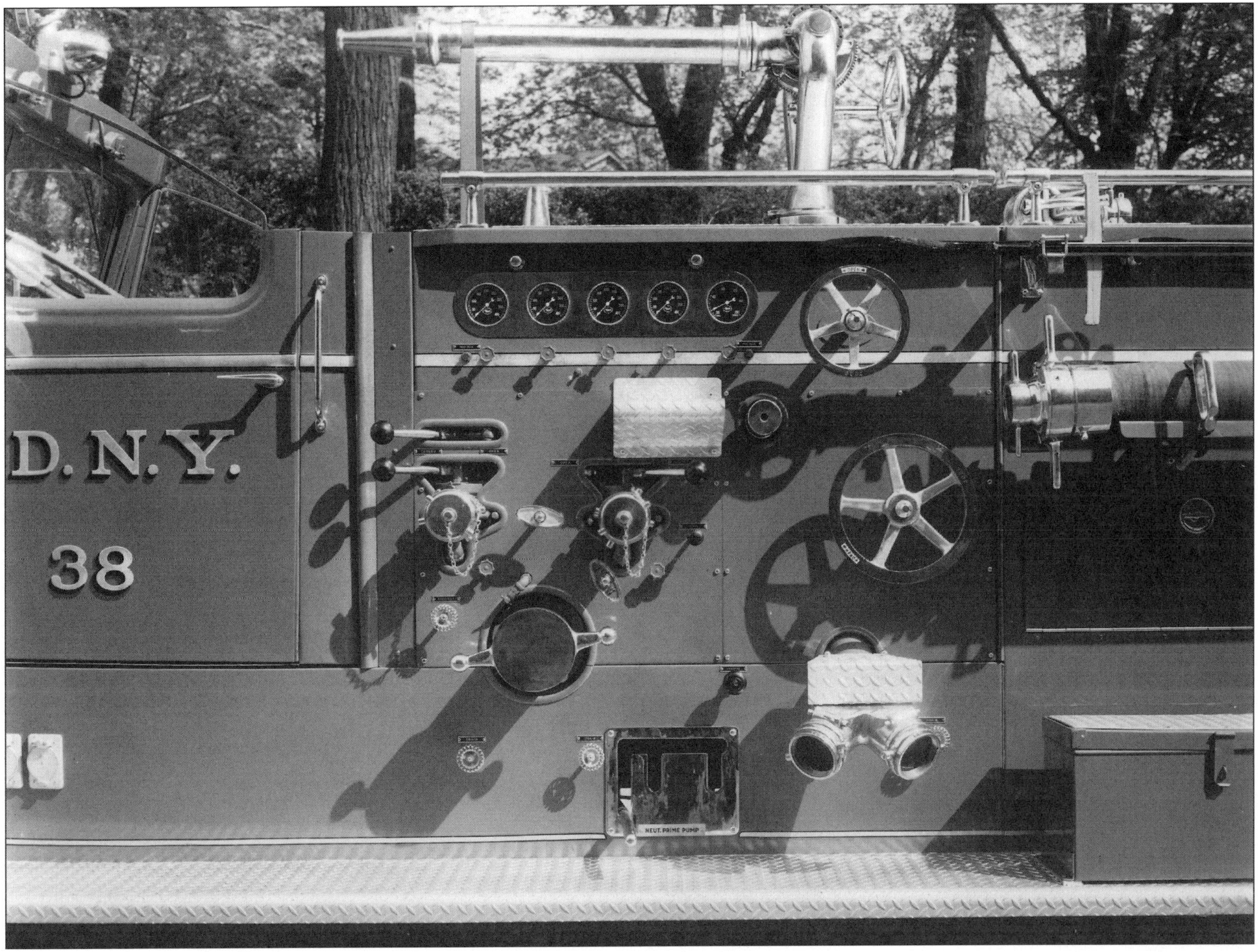

85LS1140 Pump panel details show one compound gauge and four pressure gauges. A Siamese connection is provided on each side of the rig to supply the deck pipe. During the time these rigs were in service, it was not uncommon to see a dozen or more L model Macks pumping at large New York City fires.

85LS1163 1947 Another "war wagon" was this hose and deluge unit delivered to Boston, Massachusetts, equipped with two large deck guns. The chrome and full striping made for an extremely handsome rig.

85LSl163 This rear view shows the 2,000GPM capacity deck pipes which can be supplied by four, three-way connections, two on each side. Two foam type extinguishers are mounted on the rear step.

85LS1207 1948 This sedan cab pumper for Newport, Delaware, is equipped with larger than standard ladders.

85LS1207 The right seat base normally displayed both the Mack and Hale model and serial number plates. The shift position plate on the floor shows this unit is equipped with the standard four-speed transmission.

85LS1211 1947 As you might expect, most of Allentown's fleet was a hometown product. This closed cab Quad is equipped with extra pressure gauges and top mounted flood lights.

85LS1224 1948 This cab shot shows optional air brakes and wipers, and a five-speed direct transmission.

85LS1260 1948 Creeskill, New Jersey, received this sleek Quad in the popular semi cab configuration.

85LS1260 Before the days of automatic tire chains, sanders were sometimes specified for slippery road conditions. The sand container is shown inside the compartment, with rubber tubes positioned in front of each dual tire.

85LS1260 Shown in this rear shot is the large hose bed, dual coat and boot rails, the booster line, the rear 2 1/2" connection, and concealed hard suction mountings.

85LS1002 This pump panel shot shows the standard arrangement of tachometer and intake and discharge gauges; three discharge gate handles; the transfer valve; the Mack governor; the pump shift lever; the throttle control; and the pump performance plate.

85LS1281 1948 This sedan cab has special significance as it is now owned by the editor. It was delivered to Selinsgrove, Pennsylvania, in late 1948, and cost $15,218.75. It answered its last fire call on May 17, 1987, nearly 40 years later.

85LS1281 The major options pictured include the compartment for a Homelite generator, the 300-foot cable reel, and floodlights mounted on a plate over the forward end of the hose body.

85LS1281 The 300 gallon booster tank is equipped with a "special" booster tank filler arrangement. It quickly became known in the field as a "slop fill."

85LS1343 1949 Unusual in its day, this streamlined cab pumper was equipped with extensive side compartments. It was delivered to Eggertsville, New York. It has a roof mounted siren.

85LS1343 Because of the side compartments, the suction hose and ladders were relocated as shown. The hose bed was quite small.

85LS1418 1950 Mack re-entered the aerial ladder truck field in 1949 with Maxim ladders of 65- and 75-foot lengths. This unit was delivered to Boston, Massachusetts.

85LS1418 This straight on shot shows the ladder fully extended. The L model Mack/Maxim combination quickly gained wide acceptance.

85LS1418 Supply your own caption! What do you think the gentlemen at the end of the ladder might be saying?

85LS1422 1950 Long Island, New York, was a Mack stronghold. This Lynbrook pumper is shown in "response" mode. Typical of the time, a full load consisted of the 2 1/2-inch hose, with firefighter gear carried on the coat and boot rails, and firefighters riding on the rear step.

85LS1452 1950 An Allentown, Pennsylvania, pumper pumping in front of the old Cetronia fire station. The open hood shows the 707 motor with standard chrome valve covers and air cleaner. This dual ignition 707 cubic inch motor developed 225 horsepower at 2,500rpm and developed 578ft-lbs of torque at 1,100rpm.

85LS1453 1950 This 65-foot aerial ladder was delivered to Bala-Cynwyd, Pennsylvania. The Union Fire Association had the distinction of receiving the very first Mack pumping engine in 1911. This aerial saw service in several Pennsylvania and New Jersey fire departments before being purchased by a collector.

85LS1453 A full complement of wooden ground ladders was carried.

85LS1487 1950 This pumper, delivered to Ridgefield, New Jersey, is believed to be a "one of a kind" open cab, crew cab arrangement. The gauge panel and pump location, further back than normal, gives the first hint of something different.

85LS1487 This shot clearly shows the rear seat, which can be entered to the right of the front passenger seat. I have not been able to determine the whereabouts of this really unique rig or if it still exists.

85LS1521 1951 This pumper, fully outfitted with hose and gear went to Emmaus, Pennsylvania, a neighboring town of Allentown. Fire apparatus production was now back in Allentown. Photographed at an airport runway.

85LS1521 This front shot shows the six chrome strip front channel bumper, and normal placement of the siren, bell, and searchlight. About 1949, a subtle styling change resulted in "fatter" front fenders. Most L model enthusiasts prefer the former highly crowned, narrower fenders.

85LS1557 1951 This handsome, streamlined cab pumper went to Reiffton, Pennsylvania, a Reading suburb. It is equipped with a side mounted booster reel and a generator compartment. In 1998, this company operates three Mack CF units and would, no doubt, still be an "all Mack" company if Mack still produced custom apparatus.

85LS1573 1951 In addition to the L model pumper shown on page 40, Riverdale, Maryland, also received this massive "Quint." This company had the distinction of later operating the only semi open cab Mack CF aerial built.

85LS1635 1952 This two-tone pumper served Hagerstown, Maryland, the home of Mack motor and transmission production since 1961.

85LS1685 1953 The latest L model apparatus photo in the Mack archives was this basically standard pumper for Chillum Adelphi, Maryland. Gold leaf striping was missing, although it may have been eventually added locally.

TYPE 95 and 19

Both the type 95 and 19 had the 707 cubic inch thermodyne motor and 1,000GPM Hale pump as standard. There were minor differences in the pump models, and in some cases, the type 19 used heavier chassis components. Multiple stage pumps were frequently specified on type 19 units.

The 95 and 19LSs were both produced from 1940 through 1954 with 351 95LSs built and 109 19LSs built. The first three 95LSs went to California. Palm Springs received the first on August 30, 1940. The final 95LS, serial number 95LS1362, was delivered to Lyndhurst, Ohio, on October 27, 1954.

The first 19LS went to Niagara Falls, New York, on July 2, 1940, with the next eight going to New York City. The last 19LS delivery, serial number l9LS1114, was to Portsmouth, Ohio, on July 15, 1954.

95LS1005 1941 The fifth type 95 built went to St. Louis, Missouri. Like the first types 80 and 85, it was of the open cab style.

95LS1005 The "squirrel tail" suction hose arrangement was popular during this era. The purpose was to facilitate a fast hookup, as only a single connection to a hydrant or placement in a drafting site, was required.

95LS1057 1946 This postwar sedan cab for Orlando, Florida, lacks chrome except for the hood strips, bulldog, and radiator shell emblem.

95LS1069 1947 One of Mack's West Coast customers was the Tacoma, Washington, Fire Department, which received several closed cab pumpers. The standard striping, as shown, was quite extensive.

95LS1069 The standard gauge and pump panel is shown, except for the special three-way intake with two gated 2 1/2-inch inlets.

95LS1072 1947 Open cab pumper for Hammond, Indiana, is shown in the paint shop at the Long Island City plant.

95LS1072 Unusual features are the overhead ladder rack and the bumper mounted turn signals. At this point, neither turn signals nor rear view mirrors were mandatory.

95LS1086 1948 Chicago, Illinois, received many Mack sedan cabs over the years. This big, two-tone four-door cab is equipped with disc wheels.

95LS1086 This front cab shot shows spotlights in place of a searchlight, hydraulic brakes, and round rear view mirrors. The lever to the left of the driver seat is to shift from road to pump mode.

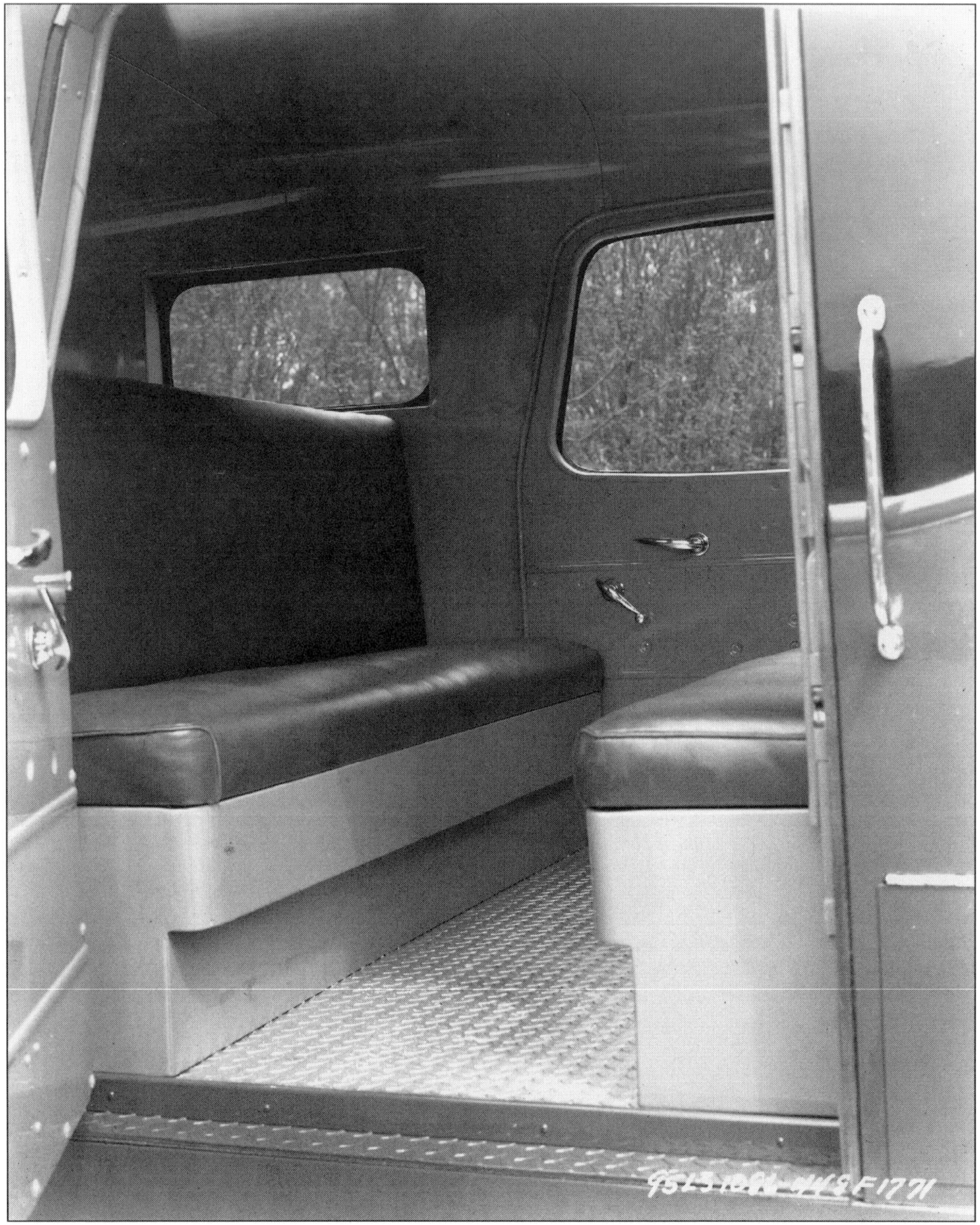

95LS1086 Don't think that nine-man cabs are something new, although three-seat L model sedan cabs were rare. I've been told the firefighters still preferred to ride the rear step in this era.

95LS1128 1949 Memphis, Tennessee, was a long-time user of sedan cabs and "squirrel tail" suction hose arrangements.

95LSl128 Because of the suction hose arrangement, it was necessary to mount the bell and siren in the location shown. The truck in the background is a "little brother" E model.

95LS1128 The gate valve on the already connected inlet hose allowed the pump operator to supply booster water until the hydrant connection was completed.

95LS1141 1949 The Mars light and bell mounting are long-time Chicago trademarks. This pumper is equipped with a left side mirror, a right side spotlight, and bumper mounted red lights.

1949 This "fleet order" for Chicago awaits delivery from the Mack Chicago Branch which was then located at 3301 S. Wentworth Avenue.

95LS1197 1950 This semi cab pumper for Hempstead, New York, is shown on the Long Island City plant factory floor. The American flag is prominent in the background.

95LSl275 1952 This semi cab for East Rockaway, New York sports a chrome bumper, triple coat and boot rails, fender scrolls, and a side mounted booster reel.

95LS1275 Dual cowl mounted sirens and red lights were frequently specified for that little extra "pizzazz."

95LS1305 1953 Quads were popular in this era. This closed cab job for Wilmington, Ohio, has full side compartments, which gained acceptance in the 1950s.

95LS1316 1953 This two-tone pumper for Reistertown, Maryland, has individual discharge gauges, an extra length of hard suction hose, and an extended wheelbase.

An order for eight 19LS hose wagons for New York City await completion in this Allentown plant photo from 1940.

19LS1007/14 1940 This FDNY order of eight hose wagons were some of the very first L model deliveries. They cost $7,327 each. A Siamese connection on both sides fed the 2,000GPM Morse deck pipe. The poor condition of New York streets, even back then, is indicated by the spare tire carried at the rear.

19LS1007/14 Unusual or unique FDNY specs included single rear tires, rear "subway straps," disc wheels, and a Mars light. Not shown are the Buckeye exhaust whistle and H.O.R. belt driven siren.

19LS1038 1944 Richmond, Virginia, took delivery of 52 Macks from 1915 through 1977. The tractor shown here was delivered in 1944 along with the early American LaFrance trailer, which was refurbished by Mack at the Long Island City plant.

Although the serial number is not listed, this 1945 Pirsch factory photo shows an L model tractor with a Pirsch tillered aerial in tow. Many departments, besides Richmond, liked the combination of a Mack chassis or tractor teamed up with a Pirsch aerial.

9LS1047 1945 This semi cab tractor for Medford, Massachusetts, has a minimum amount of chrome.

19LS1047 The fifth wheel for a tillered aerial differs considerably from a highway rig.

19LS1048 1945 This open cab pumper with a big deck pipe was delivered to Scotia, New York. Options included a bumper mounted red light, a left spotlight, and gated 2 1/2-inch pump inlets.

19LS1048 This plant shot shows the mounted pump and the plumbing for the deck pipe.

19LS1031 1942 This open cab pumper was delivered to Stockton, California. Most West Coast customers preferred disc rather than spoke wheels.

19LS1063 1947 The capitol of Pennsylvania, Harrisburg, was a "Mack town" for many years. Three L model pumpers were delivered, including this 1947 unit, which has a 1,000GPM pump, and a 200 gallon tank. It cost $12,335. It is now owned by the fire department historian.

19L81071 1948 This sedan cab for Mt. Clemens, Michigan, features dual searchlights, a Mars roof light, and individual pressure gauges.

19LS1079 1949 This two-tone tractor for Baltimore, Maryland, is especially "spiffy" with fender scrolls and dual searchlights.

TYPE 125

The type 125LS was produced from 1940 through 1954 with 46 built. The first l25LS was delivered to Menasha, Wisconsin, on April 30, 1940. The next three deliveries went to Los Angeles County in 1942. The last 125LS, serial number 125LSl049, was delivered to Long Beach, California, on July 7, 1954.

The 125LS also used the 707 thermodyne motor, but the standard pump was of 1,250GPM capacity, which was a large pump at the time, normally only specified by the larger departments.

125LS1001 1940 The very first 125LS built was this semi cab pumper for Menasha, Wisconsin.

125LS1001 A standard pump panel is shown. To pump, the driver needed to switch the lever in the cab from "road" to "pump"; put the transmission in direct gear; and move the pump panel lever from "neutral" to "pump," or to "prime" if necessary to draft from a static water source. Once primed, the lever would then be moved to "pump."

125LS1001 This top view shows the early car type bumper and non-sealed beam headlights.

125LS1040 1952 This semi cab pumper for Fort Myers, Florida, has few options.

125LS1043 1953 This open cab pumper for Fullerton, California, was no doubt to be finished at the destination. The make and location of the siren was non-standard.

TYPE 21

The type 21 was first used in the early B series, and then in the E series beginning in 1938. The L series type 21 was produced from 1941 through 1953, with 80 built. The first 22 21LSs were delivered to New York City beginning in 1941. The last three units, serial numbers 2lLS1100, 01, and 02, were delivered in 1953 to Los Angeles, California.

The type 21s were the “heavy artillery,” with pumps of two or four stages from 1,000 to 1,500GPM capacity, with two special units being delivered with 2,000GPM pumps.

These were the only L Macks delivered without Mack motors. The higher pump capacities required a higher horsepower than the 707 Mack motor could deliver. The initial motor used was a Hercules HXE, and later units were equipped with Hall Scott motors. Both brands were 935 cubic inches, but an optional 1,090 cubic inch Hall Scott blockbuster was available.

21LS1020/31 1941 These 12 units for New York City are very similar to the type l9LS hose wagons shown on pages 97, 98, and 99. Two major differences are that these are equipped with Hercules HXE motors, and Hale ZMM 1,000GPM four-stage pumps.

21LS1020/31 These big pumpers cost $13,208 each. They entered service in late 1941. An interesting and unexplained phenomenon is that the first New York LSs had closed cabs; the middle batch had semi cabs; and the final LSs delivered in 1954 were completely open.

21LS1020/31 The pump panel for this size pumper is rather stark. The lack of a pump panel tachometer is unexplained. In an unusual arrangement, the pressure gauges for the right side discharges are mounted on the right side.

21LSl042 1946 This Monterey Park, California, pumper was the second built after the switch from Hercules to Hall Scott motors.

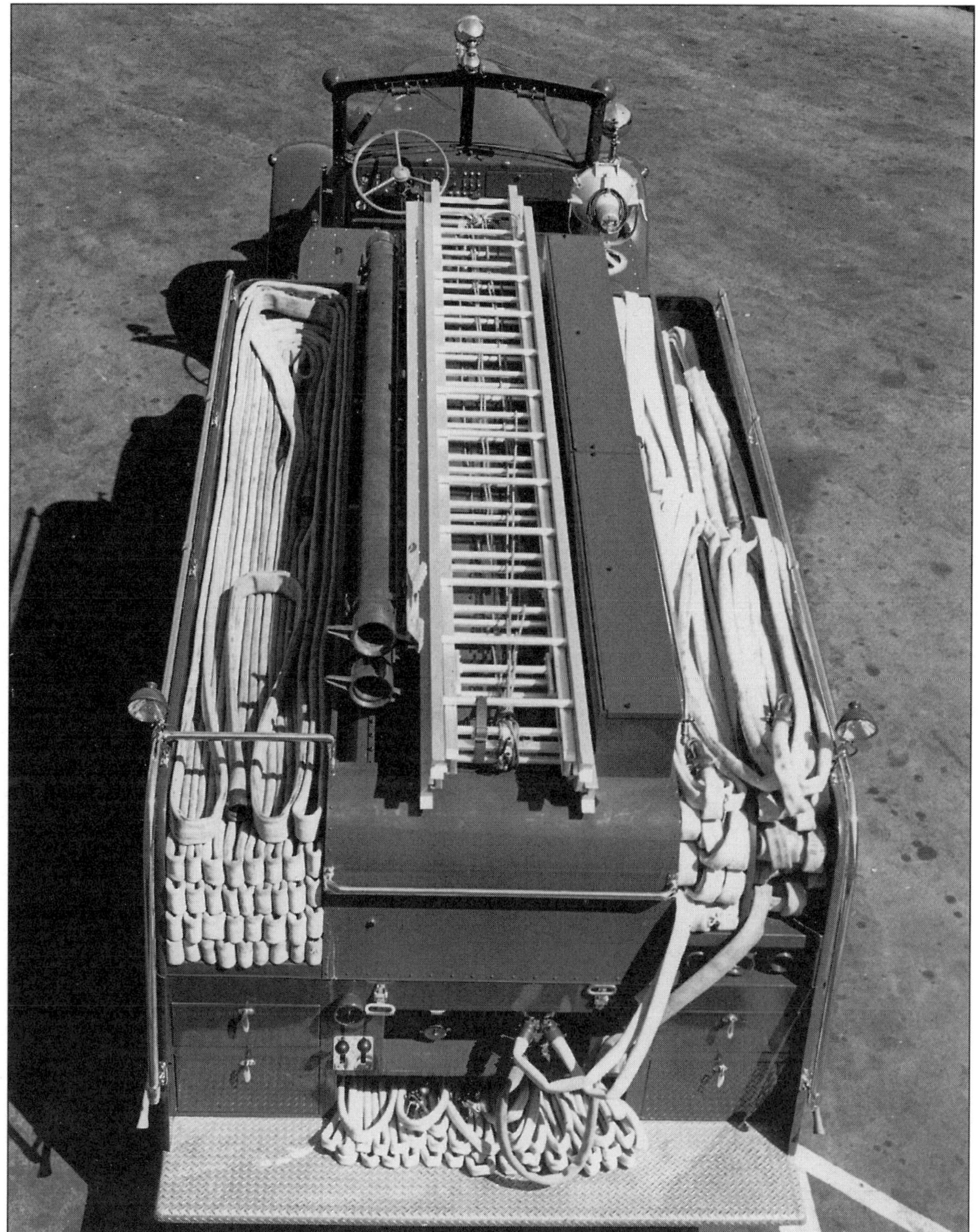

21LS1042 In all probability, this unique pump panel control and body were completed by a West Coast builder.

21LS1044 1948 This typical looking West Coast rig was delivered to South Pasadena, California. The pump connection through the front bumper was rare.

21LSl045/46 1948 Although they did not acquire the fame of the 1965 Mack Super Pumper system, this rig and a sister rig were Mack's first "super pumpers." After World War II, Chief Lockhart of the Minneapolis Fire Dept. worked with the Waterous Pump Co. to develop a large capacity pump of 2,000GPM. In 1948 the city accepted delivery of two Mack 2,000GPM pumpers, and two American LaFrance 1,500GPM units, thus allowing Minneapolis to reach the goal of running six to eight hose lines from a single pumper. These Macks were in use until 1974 and were the largest single pump machines of their day. They were powered by Hall Scott 400 motors, which had a displacement of 1,090 cubic inches, and developed 290 horsepower at 1,800rpm, and 930ft-lbs of torque at 1,300rpm.

2lLSl07l 1949 This four-door sedan for Memphis, Tennessee, was equipped with a Hall Scott 400 motor.

21LS1071 The similarities of the pump panel and general equipment location is almost identical to the setup of 1957 B model pumpers delivered to the department. (See pages 60 and 61 of "Mack Model B Fire Trucks 1954 - 1966 Photo Archive".)

2LS1079 1949 This "beefy" Quad for San Diego, California, was powered by a Hall Scott 480 motor of 935 cubic inch displacement.

21LSl102 1952 This semi cab pumper was shipped to the Los Angeles branch for final assembly. A Hall Scott 480 motor can be seen in the open motor compartment. Final delivery was to the Los Angeles, California, Fire Department. The Mack plant shown in the background is Allentown Plant 5A.

The Iconografix Photo Archive Series includes:

AMERICAN CULTURE

AMERICAN SERVICE STATIONS 1935-1943	ISBN 1-882256-27-1
COCA-COLA: A HISTORY IN PHOTOGRAPHS 1930-1969	ISBN 1-882256-46-8
COCA-COLA: ITS VEHICLES IN PHOTOGRAPHS 1930-1969	ISBN 1-882256-47-6
PHILLIPS 66 1945-1954	ISBN 1-882256-42-5

AUTOMOTIVE

FERRARI PININFARINA 1952-1996	ISBN 1-882256-65-4
GT40	ISBN 1-882256-64-6
IMPERIAL 1955-1963	ISBN 1-882256-22-0
IMPERIAL 1964-1968	ISBN 1-882256-23-9
LE MANS 1950: THE BRIGGS CUNNINGHAM CAMPAIGN	ISBN 1-882256-21-2
LINCOLN MOTOR CARS 1920-1942	ISBN 1-882256-57-3
LINCOLN MOTOR CARS 1946-1960	ISBN 1-882256-58-1
MG 1945-1964	ISBN 1-882256-52-2
MG 1965-1980	ISBN 1-882256-53-0
PACKARD MOTOR CARS 1935-1942	ISBN 1-882256-44-1
PACKARD MOTOR CARS 1946-1958	ISBN 1-882256-45-X
SEBRING 12-HOUR RACE 1970	ISBN 1-882256-20-4
STUDEBAKER 1933-1942	ISBN 1-882256-24-7
STUDEBAKER 1946-1958	ISBN 1-882256-25-5
VANDERBILT CUP RACE 1936 & 1937	ISBN 1-882256-66-2

TRACTORS AND CONSTRUCTION EQUIPMENT

CASE TRACTORS 1912-1959	ISBN 1-882256-32-8
CATERPILLAR MILITARY TRACTORS VOLUME 1	ISBN 1-882256-16-6
CATERPILLAR MILITARY TRACTORS VOLUME 2	ISBN 1-882256-17-4
CATERPILLAR SIXTY	ISBN 1-882256-05-0
CLETRAC AND OLIVER CRAWLERS	ISBN 1-882256-43-3
ERIE SHOVEL	ISBN 1-882256-69-7
FARMALL CUB	ISBN 1-882256-71-9
FARMALL F-SERIES	ISBN 1-882256-02-6
FARMALL MODEL H	ISBN 1-882256-03-4
FARMALL MODEL M	ISBN 1-882256-15-8
FARMALL REGULAR	ISBN 1-882256-14-X
FARMALL SUPER SERIES	ISBN 1-882256-49-2
FORDSON 1917-1928	ISBN 1-882256-33-6
HART-PARR	ISBN 1-882256-08-5
HOLT TRACTORS	ISBN 1-882256-10-7
INTERNATIONAL TRACTRACTOR	ISBN 1-882256-48-4
INTERNATIONAL TD CRAWLERS 1933-1962	ISBN 1-882256-72-7
JOHN DEERE MODEL A	ISBN 1-882256-12-3
JOHN DEERE MODEL B	ISBN 1-882256-01-8
JOHN DEERE MODEL D	ISBN 1-882256-00-X
JOHN DEERE 30 SERIES	ISBN 1-882256-13-1
MINNEAPOLIS-MOLINE U-SERIES	ISBN 1-882256-07-7
OLIVER TRACTORS	ISBN 1-882256-09-3
RUSSELL GRADERS	ISBN 1-882256-11-5
TWIN CITY TRACTOR	ISBN 1-882256-06-9

RAILWAYS

CHICAGO, ST. PAUL, MINNEAPOLIS & OMAHA RAILWAY 1880-1940	ISBN 1-882256-67-0
CHICAGO&NORTH WESTERN RAILWAY 1975-1995	ISBN 1-882256-76-X
GREAT NORTHERN RAILWAY 1945-1970	ISBN 1-882256-56-5
GREAT NORTHERN RAILWAY 1945-1970 VOLUME 2	ISBN 1-882256-79-4
MILWAUKEE ROAD 1850-1960	ISBN 1-882256-61-1
SOO LINE 1975-1992	ISBN 1-882256-68-9
WISCONSIN CENTRAL LIMITED 1987-1996	ISBN 1-882256-75-1
WISCONSIN CENTRAL RAILWAY 1871-1909	ISBN 1-882256-78-6

TRUCKS

AMERICAN LaFRANCE 700 SERIES 1945-1952	ISBN 1-882256-90-5
BEVERAGE TRUCKS 1910-1975	ISBN 1-882256-60-3
*BROCKWAY TRUCKS 1948-1961**	ISBN 1-882256-55-7
DODGE TRUCKS 1929-1947	ISBN 1-882256-36-0
DODGE TRUCKS 1948-1960	ISBN 1-882256-37-9
DODGE POWER WAGONS 1940-1980	ISBN 1-882256-89-1
LOGGING TRUCKS 1915-1970	ISBN 1-882256-59-X
*MACK® MODEL AB**	ISBN 1-882256-18-2
*MACK AP SUPER-DUTY TRUCKS 1926-1938**	ISBN 1-882256-54-9
*MACK MODEL B 1953-1966 VOLUME 1**	ISBN 1-882256-19-0
*MACK MODEL B 1953-1966 VOLUME 2**	ISBN 1-882256-34-4
*MACK EB-EC-ED-EE-EF-EG-DE 1936-1951**	ISBN 1-882256-29-8
*MACK EH-EJ-EM-EQ-ER-ES 1936-1950**	ISBN 1-882256-39-5
*MACK FC-FCSW-NW 1936-1947**	ISBN 1-882256-28-X
*MACK FG-FH-FJ-FK-FN-FP-FT-FW 1937-1950**	ISBN 1-882256-35-2
*MACK LF-LH-LJ-LM-LT 1940-1956 **	ISBN 1-882256-38-7
*MACK MODEL B FIRE TRUCKS 1954-1966**	ISBN 1-882256-62-X
*MACK MODEL CF FIRE TRUCKS 1967-1981**	ISBN 1-882256-63-8
*MACK MODEL L FIRE TRUCKS 1940-1954**	ISBN 1-882256-86-7
STUDEBAKER TRUCKS 1927-1940	ISBN 1-882256-40-9
STUDEBAKER TRUCKS 1941-1964	ISBN 1-882256-41-7
WHITE TRUCKS 1900-1937	ISBN 1-882256-80-8

The Iconografix Photo Gallery Series includes:

CATERPILLAR PHOTO GALLERY	ISBN 1-882256-70-0
*MACK TRUCKS PHOTO GALLERY**	ISBN 1-882256-88-3

The Iconografix Photo Album Series includes:

CADILLAC 1948-1964	ISBN 1-882256-83-2
CORVETTE PROTOTYPES & SHOW CARS	ISBN 1-882256-77-8
DODGE PICKUPS 1939-1978	ISBN 1-882256-82-4
FIRE CHIEF CARS 1900-1997	ISBN 1-882256-87-5
LOLA RACE CARS 1962-1990	ISBN 1-882256-73-5
LOTUS RACE CARS 1961-1994	ISBN 1-882256-84-0
McLAREN RACE CARS 1965-1996	ISBN 1-882256-74-3
PORSCHE 356 1948-1965	ISBN 1-882256-85-9

All Iconografix books are available from direct mail specialty book dealers and bookstores worldwide, or can be ordered from the publisher. For book trade and distribution information or to add your name to our mailing list contact

Iconografix
PO Box 446/BK
Hudson, Wisconsin, 54016

Telephone: (715) 381-9755
USA (800) 289-3504
Fax: (715) 381-9756